Alim Nisa
Sajawal Ashraf
Fatima Sajjad

Neem tree: A Sacred Gift of Nature

Alim Nisa
Sajawal Ashraf
Fatima Sajjad

Neem tree: A Sacred Gift of Nature

Majestic Neem Tree

Noor Publishing

Imprint
Any brand names and product names mentioned in this book are subject to trademark, brand or patent protection and are trademarks or registered trademarks of their respective holders. The use of brand names, product names, common names, trade names, product descriptions etc. even without a particular marking in this work is in no way to be construed to mean that such names may be regarded as unrestricted in respect of trademark and brand protection legislation and could thus be used by anyone.

Cover image: www.ingimage.com

Publisher:
Noor Publishing
is a trademark of
Dodo Books Indian Ocean Ltd., member of the OmniScriptum S.R.L Publishing group
str. A.Russo 15, of. 61, Chisinau-2068, Republic of Moldova Europe
Printed at: see last page
ISBN: 978-620-3-85934-8

<u>ARTICLE</u>

ON

<u>*NEEM TREE: A SACRED GIFT OF NATURE*</u>

WRITTEN

BY

<u>MRS. ALIM-UN-NISA</u>
PRINCIPAL SCIENTIFIC OFFICER

<u>M SAJAWAL ASHRAF</u>

<u>FATIMA SAJJAD</u>

<u>FOOD AND BIOTECHNOLOGY RESEARCH CENTER</u>
<u>PCSIR LABORATORIES COMPLEX, FEROZEPUR ROAD, LAHORE-54600</u>

Table of Contents

Contents

Table of Figures

Tables

Majestic Neem Tree

Introduction

Neem or Azadirachta indica is the multifaceted medicinal plant species which is most widely distributed in sub-continental area including many states of India, Pakistan, and Sri Lanka etc. But due to its immense medicinal and economical potential, it has transferred to both tropical and subtropical zones where they start the cultivation of Neem. Neem tree possesses some majestic properties and are considered as rich source of limonoids, some essential amino acids, and several fatty acids. Moreover, Neem tree is considered as the heaven tree due to its enormous beneficial aspects including its products such as Neem oil used as for various medicinal purposes, Neem cake, and Neem seed kernel extract have extensive application in agriculture or agro-forestry.

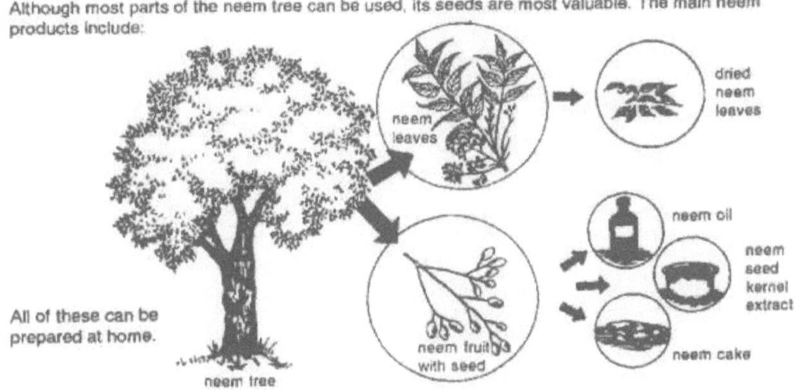

Figure 1: Majestic Neem Tree with its products.

What is Neem?

In botanical terminology, Neem is known as *Azadirachta indica* basically this word has been derived from Farsi word known as *"Azad diraklzat-I-Hind"* which means the noble or free tree of India because this tree is free from diseases and pests and do not harm the environment. **(Ahmed, 1995).** Neem or Azadirachta indica is known as one of the most beneficial species for human beings. Commonly, this neem tree is called as Margosa or Indian Lilac. Neem is considered to be the multifarious tree in the tropics with huge diversity, versatility, and immense potential. Neem belongs to the family Meliaceae and is considered as a botanical cousin of mahogany.

Figure 2: Classification of Neem Tree.

Neem is one of the fastest evergreen growing tree that can reaches a height of about 15-20 meter, but in very favorable conditions, neem tree reaches approximately up to 30-35 meter. Neem tree branches are widely distributed or spread in soil ground. And their dense crown is oval or roundish, in old free specimens, it may reach a diameter of about 15-20 meter with a relatively short trunk having a girth of about 1.5-3.5 meter. Neem bark is usually hard scaly or fissured with whitish gray to reddish brown appearance. Their heart wood is reddish in color while their sap wood is grayish white in appearance. Talking about their root system, neem tree comprises of well-developed lateral roots with strong taproot system. Neem lateral surface root may reaches up to 18 meter or above. Neem tree roots are usually associated with VAM known as Vesicular arbuscular mycorrhiza, due to this reason neem is highly considered vesicular arbuscular mycorrhizal dependent plant species. Talking about Neem tree leaves, they are usually unpaired, pinnate which are about 20-30 cm long, and while dark green leaflets reaches about 3-8 cm length are usually about 31 in numbers. Their petioles are short and their terminal leaf is usually not present and the leaflet shape is usually less or more asymmetric in nature. Neem tree flowers are white which are arranged in axillary form having drooping panicles which may attain the length of about 25 cm. Neem fruits are olive like drupes which have greenish appearance in younger stage and yellowish green to reddish appearance when they become mature, their fruits might vary in shape or length usually from elongated ovea to roundish. The inner surface of fruit is very fibrous and contains bitter sweet pulp which is about 0.3-0.5 cm thick, while the outer part of fruit is exocarp which is thin and elastic. Neem tree is usually become fully productive in almost 10 years, while at the age of 3-5 years, it starts to produces fruits. It is estimated that it can produces almost 50 kg of fruits per year and can survive up to 2 or more centuries.

Figure 3: Majestic Neem Tree

Geographic Distribution

It is thought that neem is originated in Assam and Burma, but its exact origin is unknown and is considered native of Indian sub-continents. Neem is widely distributed in both tropical and sub-tropical zones of Asia, Australia, America, Africa and the South Pacific zones, but the neem tree is the native of Indian sub-continent. As, neem is native tree of India, it is widely spread in many states of India. While in Australia, neem is introduced in country about 65-70 years ago. Neem tree in Myanmar is most common in central parts of country. Neem tree exists in only in eastern parts of Indonesia. Similarly, neem is introduced to Philippines about in the last century. Neem exists in the Fiji Islands of the South Pacific zones. Neem is cultivated in subtropical islands of southern china. While in Nepal, they are present in southern low-lying areas. Neem tree in Sri Lanka is distributed in northern parts of islands. While in Iraq, neem is cultivated in Arabian Peninsula. Neem tree is also cultivated in Abu Dhabi and in Qatar by using desalted seawater in different zones. In Makkah Arafat plains, large-scale neem tree cultivation is performed for providing shade for pilgrims at Arafat plains.

Neem tree due to its immense diversity possesses different common names according to sites of location.

Table 1: Common names for Majestic Neem Tree

Common Names for Neem Tree	
English	Neem, Indian lilac
Hindi	Neem, Nimb
Urdu	Nim, Neem
Sanskrit	Nimba, Nimbou, Arishtha (reliever of sickness)
Indonesia	Mindi
German	Niembaum
French	Azadira d'Inde, Margousier, Azidarac, Azadira
Portuguese	Margosa (Goa)
Spanish	Margosa, Nim
Tamil	Vembu, Veppan
Sri Lanka	Kohomba
Singapore	Kohumba, Nimba
Nigeria	Dongoyaro
Kiswahili	Mwarubaini (Muarobaini)

Propagation of Neem Tree

Neem tree can easily be propagated both by vegetative or by sexually using seedlings, saplings, seed, or using tissue cultures etc. We simply uses seed for the propagation of neem tree by planted it to the targeted site or also by transplanting the seedlings from plant sources. Although, seed propagation is simple and easy way for the cultivation of neem but it has been reported that neem seed viability does not stand for longer period that is, it is considered that neem seed are generally viable for germination up to 2-6 months only. But in recent advances to preservation of neem seed in France has indicated that we can preserve the neem seed up to 5 years with about 42 percent germinate capacity of neem seed.

Growth of Neem Tree

Neem tree is considered to grow almost anywhere as it is renowned for good growth in low land tropics or in dry infertile soils. However, the most suitable site or area for neem tree growth is that at which the annual rainfalls is in between 400-1200 millimeters. Neem tree does not withstand freezing or cold temperatures but in hot conditions neem tree thrives to these conditions where neem shade temperature crosses the 50°C. Neem tree usually performs better activities as compared to other plants where soil surface is not suitable such as sterile, stony or hard surfaces. Moreover, neem tree can also grow best on acidic soils and in basic soils it shed their leaves to neutralize the pH for better growth. However, neem tree unable to withstand against waterlogged conditions and quickly dies in these conditions.

Ecology

Neem tree is famous for his survival in drought or desiccated conditions. Normally, neem tree grows best in those region where annual rainfall is between (16-47 in), but also grows best in those regions where annual rainfall is below (16 in) but in these region neem tree depends on the ground water levels. Also neem tree can be cultivated in different types of soil, but most favorable soil type is sandy well drained deep soil. Neem tree is also tagged as tropical or subtropical tree and these tree can tolerate temperature ranges between 21-32°C, but cannot tolerate that temperature which is below than 5°C. Traditionally, neem tree were used for shade around temples, streets, and for other public buildings, as these neem tree are not sensitive about water quality and can tolerate bad quality of water at a great limit. Also, these neem tree have been planted especially in very dry areas or large lands in order to avoid dry and humid conditions.

Chemistry of Neem Tree

In 1919, the Indian Chemists, firstly investigated the chemistry of neem tree by isolating the acidic principle named as Margosic acid present in neem. But, in 1942, the real chemical research work has been done by isolating the three main active components such as nimbidin, nimbin, and last one is nimbiene. While in 1963, an Indian scientist studied the extensive chemistry of neem and several compounds or elements from neem has been isolated as well as characterized. From an extensive research work on neem has proved that almost all the constituents or compounds possesses same chemical structures and also derivatives of tetra-cyclic terpenes. Moreover, the compounds derived from neem also possesses wide applications in biological activity such as possesses properties of antifeedants or pesticides.

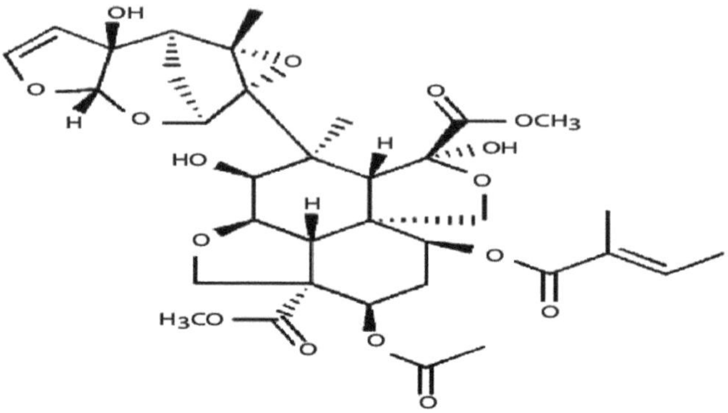

Figure 4: Chemical Structure of Azadirachtin Neem Tree.

✓ Leaves

Neem tree leaves are usually elongated to oblong in appearance and are medium to large in size with approximately 20-40 cm in length. Neem leaves possesses grassy aroma and taste and are extremely bitter in nature and usually grow along the branches of neem trees. Neem tree leaves are available all year long. Neem tree leaves comprises carbohydrates of about 23%, proteins of about 7.1%, and certain minerals like carotene, phosphorous, calcium, and vitamin C etc. Moreover, they also comprises of various fatty acids like elcosanic tetradecanoic etc. and also some amino acids like aspartic acid, glutamic acid, glutamine, alanine, tyrosine etc.

Neem leaves also contains flavonoids known as quercetin which possesses antifungal and antibacterial properties and used in the manufacturing of various medicine as a cure for treating sores or other allergic reactions. Moreover, neem tree leaves also comprises of Liminoids including the derivatives of nimbin, these are helpful for controlling mosquitoes and house flies number by producing some intermediates as these neem tree leaves possesses some mutagenic properties. Neem tree leaves extracts yields essential neem oil which show excellent antifungal properties against various fungi and inhibit their growth and activities when introduced in vitro, the researchers or scientists are working or studying the neem tree extracts oils for large scale up due to enormous potential of neem tree leaves.

Table 2: Chemical compositions present in Neem Tree Leaves

Chemical Composition	Units as Prescribed in parenthesis
Moisture	59.4%
Fat	1%

Carbohydrate	22.9%
Carotene	1998 µg/100g
Glutamic acid	73.4 mg/100g
Aspartic acid	15.51 mg/100g
Niacin	1.51 mg/100g
Calcium	511 mg/100g
Phosphorus	80 mg/100g
Iron	17 mg/100g
Tyrosine	31.51 mg/100g
Proteins	7.2 %
Vitamin C	218 mg/100g
Alanine	6.41 mg/100g
Minerals	3.4%
Fiber	6.25%
Glutamine	1 mg/100g

✓ Flowers

Neem flowers are usually white in appearance arises from petiole or stem junction present in neem tree and they are present in the form of panicles means in the form of clusters of about 25 cm length. A single neem flower is about 8-11 mm wider and 5-6 mm long in general appearance. Flowers of neem comprises of various flavonoids also with nimbosterol such as melicitrin etc. Just like leaves of neem tree, the flowers of neem also contains some fatty acids such as

- ✓ Palmitic acid of about 13.6%
- ✓ Stearic acid of about 8.2%
- ✓ Linoleic acid of about 8%
- ✓ Oleic acid of about 6.5%
- ✓ Behenic acid of about 0.7%
- ✓ Arachidic acid of about 0.7%

Moreover, the pollen of neem flowers also contains various amino acids such as phenylalanine, arginine, glutamic acid, methionine, histidine etc.

Figure 5: Flowers of Neem tree

✓ Bark

Neem bark is usually hard scaly or fissured with whitish gray to reddish brown appearance. Their heart wood is reddish in color while their sap wood is grayish white in appearance. The bark of neem tree contains various anti-inflammatory and anti-tumor polysaccharides comprises of glucose, fructose, and arabinose. Moreover, from the bark of neem tree, we have isolated various di-terpenoids from the root and stem barks of neem tree such as nimbidol, nimbolicin, nimbione, and margocin etc. Stem bark of neem tree also contains tannins of about 12-16% and also non-tannins of about 8-11%.

Besides stem bark, the heartwood of neem tree comprises potassium, calcium and iron salts which on destructive distillation provides pyroligeneous acids of about 38.4% and charcoal of about 30%. Neem wood and its extracts contains lignin of about 14.63%, cellulose and hemicellulose of about 14%, beta-sitosterol etc.

✓ Gum

Neem tree releases a gum which produces D-galactose, L-arabinose, D-glucuronic acid, and L-fructose. Moreover, the older neem tree releases a sap which comprises of amino acids such as glycine, arginine, alanine, praline, and aspartic acid etc. and free sugars such as mannose, glucose, xylose, and fructose. Also contains some organic acids like malonic acid, succinic acids, citric acids, and fumaric acids. Due to these reasons, neem tree gum or sap provide efficient treatment for treating skin infections and general weakness also.

✓ **Seed**

Due to high lipid content and large amount of essential compounds, the neem seeds are very effective against more than 300 different pests and the most important thing about neem seed is that they are non-toxic for humans. We have extracted the brownish yellow oil from neem kernels which contains:

- ❖ Glycerides
- ❖ Oleic acids of about 50-60%
- ❖ Stearic acid of about 14-19%
- ❖ Palmitic acid of about 13-15%
- ❖ Linoleic acid of about 8-16%
- ❖ Arachidic acid of about 1-3%

The overall composition present in neem seed may vary but approximately contains:

- ❖ Carbohydrates of about 26-50%
- ❖ Crude proteins of about 13-35%
- ❖ Crude fiber of about 8-26%
- ❖ Fats of about 2-13%

Moreover, we can extract cake from neem seeds by using 70% alcohol with hexane provides meal which we can used as feed for animals or poultry field, as the neem cake is full of essential amino acids. Also the neem cake comprises of tri-terpenoids with essential nutrients such as nitrogen, Sulphur, calcium, potassium etc.

Figure 6: Neem Tree Seeds

Phytochemicals in Neem

Neem tree leaves, stem, bark, fruit, and seeds contains diverse varieties of phytochemicals such as the leaves comprises of carotenes, quercetin, and vitamin C etc. Similarly Azadirachtin present in neem seed extracts, also the neem extracted oil comprise of limonoids compounds as phytochemicals. Neem tree due to these phytochemicals protect itself from various pests as these phytochemicals contains pesticidal ingredients.

✓ Limonoids

In neem tree, the researchers have isolated almost nine important neem limonoids and demonstrated that these limonoids have great ability of blocking the growth of insects or pests of wider diversity affecting human health and agricultural crops. Currently, the researchers are conducting different research for further exploration of new limonoids in neem tree. However, the nimbin, salannin, meliantriol, and Azadirachtin have gained immense popularity.

✓ Azadirachtin

Azadirachtin was one of the first main active agent isolated from the neem tree plant which have great potential to fight against different pests and insects. The efficiency of Azadirachtin for the killing of pests or insects is about 90%. Exploration in the course of recent years has shown that it is perhaps the most intense development controllers and taking care of obstructions at any point examined. It will repulse or decrease the taking care of numerous types of vermin creepy crawlies just as certain nematodes. Truth be told, it is strong to such an extent that a simple hint of its essence keeps a few bugs from contacting plants.

Azadirachtin is primarily like bug chemicals called "ecdysones," which control the course of transformation as the bugs pass from hatchling to pupa to grown-up. It influences the corpus cardiacum, an organ like the human pituitary, which controls the emission of chemicals. Transformation requires the cautious synchrony of a huge number and other physiological changes to be effective, and azadirachtin is by all accounts an "ecdysone blocker." It hinders the bug's creation and arrival of these imperative chemicals. Bugs then, at that point won't shed. This obviously breaks their life cycle.

By and large, neem portions contain somewhere in the range of 2 and 4 mg of azadirachtin for each gram of part. The most noteworthy figure so far revealed 9 mg for every gram was estimated in examples from Senegal.

✓ Meliantriol

Meliantriol is also one of the main inhibitor of pests and insects extracted or isolated from the neem tree is found to be effective even if the low concentrations has been taken. The exhibition of its capacity to forestall grasshoppers biting on crops was the principal logical evidence for neem's customary use for creepy crawly control on India's harvests.

✓ Salannin

Salannin is the third main tri-terpenoids which has been isolated from the neem tree. Studies show that this compound additionally effectively hinders feeding of various pests and insects, yet doesn't impact bug sheds. The transient grasshopper, California red scale, houseflies, and the Japanese bug, striped cucumber insect have been firmly dissuaded in both research facility and field tests.

✓ Nimbin and Nimbidin

Nimbin and Nimbidin are the two major compounds which have been isolated or extracted from neem tree and these two compounds possesses some antiviral properties. These compounds influence vaccinia infection, fowl pox infection, and potato virus X. They could maybe open an approach to control these and other viral sicknesses of harvests and domesticated animals.

✓ Others compounds

Besides these major compounds, some minor compounds or ingredients also work to perform functions of Anti-hormones. Research have explored that these certain minor compounds can prevent the pests and insects by inhibiting the swallowing mechanism. Such as diacetyl-azadirachtinol limonoids found from the neem tree. This fixing, confined from new natural products, gives off an impression of being just about as powerful as azadirachtin in examines against the tobacco budworm, yet it has not yet been generally tried in field practice.

History of Usage

Neem tree history is interconnected with the history of Indian civilization. Neem – the incredible restorative tree of India has developed with the human settlement all around the nation and has been a necessary piece of the Indian lifestyle for quite a long time. The Neem tree has for seemingly forever been a companion and defender of the Indian locals. For a long time Indians have believed this tree to brace their wellbeing and cure of infections. Moreover, it has been utilized for ensuring food and put away grains and as a compost and regular pesticide for our crops. Neem tree has been utilized for a far more extensive exhibit of employments than some other tree

The Neem tree (*Azadirachta indica*) was most likely India's trick of the trade. Old India was begrudged for its Cardamom, Saffron, Black Pepper, Turmeric, Silk, Sandalwood and so forth and these valued fixings were pursued and moved towards the oceans to Europe for quite a long time. The British Raj likewise neglected to get a handle on the meaning of the presence of this tree in each niche and corner of India. Maybe, on the off chance that they had thought about the awesome exhibit of employments of the Neem Tree, it would have become an overall marvel a long time back. Especially for Sub-continental Indians, the Neem tree provide many captivating perspectives. For the kids this evergreen, majestic tree was a shelter from rainfall and sunlight— they went through hours in its cooling conceal, culled the sweet ready natural product for a tidbit and fabricated tree houses, which they imparted to butterflies, birds and honey bees. This tree was the anointed one since its shade is known to be cooler than some other tree's, and furthermore, no insects and pests are in sight under this is a result of its repellant activity.

Neem tree for the women of sub-continents was the backbone of the home grown magnificence custom. Neem tree was a wellspring of medication to treat in excess of a 100 medical conditions, from skin rashes, scratches, diabetes, and especially from malarial fever. The ladies likewise utilized it to secure their put away pulses and grains as the year progressed.

Similarly, for the men of sub-continent, the neem tree gave leaf, seeds, and bark which could be changed over into compost and vermin control material. It likewise gave restorative mixtures to their dairy cattle and domesticated animals. Plus, the breeze that blew through the branches of the tree kept their homes liberated from microbes and infections and cool their houses throughout the mid-year.

For quite a long time Indians have cultivated the neem tree nearby their homes and rehearsed delicate and day by day communication with this unprecedented plant. For ladies specifically, the Neem demonstrated an important wellspring of wellbeing, cleanliness and magnificence that was unreservedly accessible. Having a shower with a decoction of neem leaves kept their skin graceful and solid. Neem leaf powder or squashed leaves fused into their face packs gave emollient and hostile to maturing activity. The sterile properties of neem tree leaves extract helped in the prevention of acne, skin infections, pimples and skin break out.

In some states of sub-continental India, it was a standard practice to apply lamp black at the edge of the eye, especially by young women as a stunner help to make eyes obvious. The normal technique utilized to make lamp black was to take an earthen light and put neem oil and a cotton wick in it. At the point when lighted, the wick freed overflowing smoke from which light dark could be gathered, by putting a metal cup containing water for cooling, some separation away from the fire. The lamp black residues was then scratched from under the cup and blended in with a little amount of mustard oil to form a paste called as Kaajal.

Neem Oil

Neem tree extracted oil possesses some magical properties through which we can prevent baldness and turning gray of hair and was utilized as against lice and hostile to dandruff treatment. When a dried neem leaf powder of about a teaspoon is blended in with a similar amount of honey and ghee was known to assist control with cleaning sensitivities.

A combination of equivalent amounts of extracted neem seed powder, rock alum and salt blended and was utilized for the maintenance of dental teeth and gums. Nimba, the extraordinary medication for the treatment of pitta and for the purification of blood.

Different studies and researches have been conducted on neem tree and their derivatives compounds, we realize today that this exceptional tree can do all that it can do on account of the sheer scope of mixtures present in neem tree. Current examination has uncovered the mystery of its viability. Its amazing antibacterial, hostile to parasitic, antiviral and disinfectant properties make it especially viable in getting anything from dandruff skin break out, dermatitis to intestinal sickness and mouth blisters to moles. Amusingly, it is this very adaptability that for such a long time has held this tree and its astounding properties back from becoming the overwhelming focus. Every buyers mindset have been molded into directs that there must be an expert answer for every issue, all together for the answer for be successful. That one neem tree can tackle such countless different issues is basically hot topic in the present market.

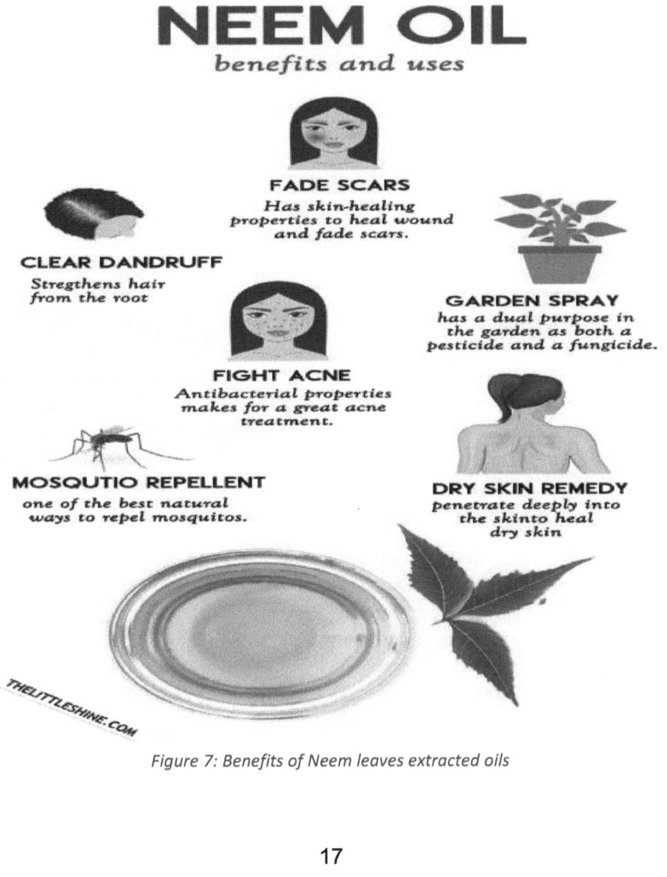

Figure 7: Benefits of Neem leaves extracted oils

Neem oil possesses exceptional properties including anti-bacterial, anti-fungal, anti-viral, and antiseptic properties. As these neem oil causes no complications as compared to the synthetic chemicals which causes allergic reactions, skin rashes, and skin diseases etc. Moreover, neem oil can be used as an anti-aging formula, for making soothing gels, face masks, face washes etc.

❖ Mosquito Repellent

Neem oil is an efficient repellent for mosquitoes and other insects and do not causes any complication when sprayed into body.

❖ Clear Dandruff

As neem oil has anti-fungal properties, neem oil is one of the best way to treat or remove the dandruff naturally. Moreover, neem oil provides relief from itchy scalp or inflammation.

❖ Relieve Dry Skin

As neem oil contains some essential elements including Vitamin E and some fatty acids which provides relieve from dry skin as it makes the skin glow and nourished.

❖ Fight Acne

Neem oil has anti-bacterial properties which are used against those bacteria which causes acne, as a result we can control or prevent the acne, scars by using this neem tree oil.

Neem and Environment

The all inescapable utilization of fabricated materials in all social statuses, be it agribusiness, clothing, protection or medical care is currently clearing way for a quest for eco-accommodating items. Globally, all people group show their trust and depend more on green innovation than any other time in recent memory since the appearance of current science. The time of reliance on engineered synthetic compounds of the early and the mid-20th century provoked blend of more current synthetic substances as a panacea for all sicknesses and infirmities. The moderate disposition of certain social orders, which relied upon normal items in inclination to the engineered, was frequently credited to idleness or backwardness.

Currently, modern societies winding up jumbled in the snare of their creation, and are willing to return to nature for cures". Neem has arranged a rebound and vows to hold the focal point of the audience in the coming years. Neem play a very vital role to protect the climate from contamination; since its in-brilliance is sanitizing with its padded peaks throwing fifty feet into the sky neem for providing healthy environment. Like different trees, it breathes out oxygen and keeps the oxygen level in the air adjusted.

Just like different trees, neem tree also provide ecological advantages, for example, flood control, diminished soil disintegration and less salination. Neem can deflect ecological emergency in India and other tropical nations as it tends to be effectively utilized for restoration of environments and waste terrains. Neem is enthusiastically suggested for reforestation of dry regions in India and tropical sites of the sub-Saharan district, Asia. Neem is incredibly valuable in metropolitan ranger service since it has striking capacity to withstand air and water contamination just as warmth. Neem likewise helps in reestablishing and keeping up with soil fruitfulness which makes it exceptionally appropriate in agro-ranger service. Neem is a characteristic asset to keep climate clean. In towns and urban areas just as on farm, it is valuable as a windbreak. As a wellspring of shade, it is astounding for parks, side of the road and so forth In view of its countless characteristics, it is a typical practice in rustic India to have a neem of tree inside the mixtures of the majority of the houses. Neem tree species is regarded as the most valuable tool in agro-forestry services.

Figure 8: The effect of Neem Tree on Earth's Environment

Neem possesses some excellent pest controlling exercises and restorative properties. The most significant ability of neem tree produced pesticides are a lot more secure contrasted with engineered pesticides. The results of the engineered pesticides are regularly not less genuine than the actual issues. They cause enviro-pollution and poses great danger to human wellbeing. As an outcome, there has been an exceptional quest for more secure pesticides.

Pesticides produced from neem tree are non-toxic and can easily be degraded into simpler products which poses no harm to our environment. In short, we can say that neem tree produce no evil results to people and creatures; they have no leftover impact on agribusiness produce. Therefore, Neem is considered as the best alternatives to conventional dangerous pesticides. Moving towards the sustainable environment by organic farming, the utilization of plant items as pesticides has obtained more prominent importance. Neem is a profoundly reasonable contender for climate well disposed, safe agribusiness advancement. Azadirachtin can be utilized in horticulture and general wellbeing as an eco-accommodating substance. Utilization of neem items for plant assurance will decrease the interest for compound pesticides and subsequently lessen the natural heap of these engineered pesticides. Moreover, these bio-pesticides are far better than synthetic pesticides, as they do not poses health problems to humans.

Environmental services rendered by Neem

During blistering summers in northern pieces of the Indian subcontinent, the temperature under the neem tree is ~10° C not exactly the encompassing temperature, it is achieved when we used almost 10 air conditioner systems worked together may not do the work as productively and financially as a totally mature neem. Rebuilding of the wellbeing of degraded soils and extreme utilization of such recovered lands through neem is another illustration of its worth as a natural panacea.

Besides the natural beauty of neem tree, it provides magnificence and peacefulness, yet additionally serve itself as a house of various valuable living beings such as bugs, birds, bumbles bee, bats, and so forth. Honeycombs set up on the neem tree are uniquely liberated from the galleria wax moth invasion. Numerous types of birds and natural product eating bats stay alive on the sweet tissue of ready organic products, while certain rodents specifically feed on the portion, affirming neem's security to warm-blooded creatures. The litter of falling leaves further develops soil fruitfulness and the natural substance. As of now, little is thought about the mycorrhizal relationship among neem and bacterial and parasitic endophytes, yet the tree is by all accounts a living microcosm.

With regards to 10 years prior, about 50,000 neem trees were planted more than 10 kms on the Plains of Arafat to give shade to Muslim explorers during hajj. The neem manor notably affects the region's microclimate, micro-flora, micro-fauna, sand soil properties, and when completely mature could give shade to 2 million pioneers (Ahmed, 1995). It is an old conviction that neem developing inside the house can keep the encompassing air clean of debasements and subsequently control ecological contamination. Additionally, hanging neem twigs on the entryway of a house is said to offer assurance against contamination and infection. If the neem tree is not cut down, than the evergreen neem tree can survive up to 200-300 years. Indeed, even a profoundly moderate gauge of the 'ecological help' delivered by the tree @ US $ 10 every month, would give an astounding worth of US $ 24,000 to 36,000 in its life time.

Neem in reforestation and agro-forestry

As already mentioned that Neem is one of the most valuable forestry species in sub-continent regions and is also becoming popular in Tropical America, the middle-east countries and in Australia. Being a hardy, multipurpose tree, it is ideal for reforestation programs and for rehabilitating degraded, semiarid and arid lands. In 1987, a severe drought in Tamil Nadu State was witnessed that neem grew luxuriantly, while other plants vegetation dried up in this severe drought.

Neem is valuable as windbreaks and in spaces of low precipitation and high wind speed. In the Majjia Valley in Niger, more than 500 km of windbreaks involved two fold columns of neem trees have been planted to secure millet crops which brought about a 20% increment in grain yield (Benge, 1989). Neem, windbreaks on a more limited size have likewise been developed along sisal manors in beach front Kenya. Enormous scope planting of neem has been started in the Kwimba Afforestation Scheme in Tanzania. In nations from Somalia to Mauritiania, neem has been utilized for ending the spread of the Sahara desert. Likewise, neem is a favored tree along roads, in business sectors, and close estates in view of the shade it gives. In any case, neem is best planted in blended stands. Neem has every one of the great characters for different social ranger service programs. Neem is excellent tree for silvi-pastoral framework including creation of scavenge grasses and vegetables. In any case, as indicated by certain reports (Radwanski and Wickens, 1981), neem can't be developed among rural yields because of its forceful propensity. Others say that neem can be planted in blend with natural product societies and yields like sesame, cotton, hemp, peanuts, beans, sorghum, cassava, and so on, especially when neem trees are as yet youthful. The neem tree can be hacked to lessen concealing and to give grain and mulch. Late advances in tissue culture and biotechnology should make it conceivable to choose neem aggregates with beneficial tallness and height for use in intercropping and different agro-ranger service frameworks. The allelopathy impacts of neem on crops, assuming any, should be researched.

Figure 9: Neem in reforestation and in agro-forestry.

Biomass production and utilization

Completely mature neem trees yield between 10 to 100 tons of dried biomass/ha, contingent upon precipitation, site qualities, dividing, ecotype or genotype. Leaves include about half of the biomass; foods grown from the ground comprise one-quarter each. Further developed administration of neem stands can yield harvests of about 12.5 cubic meter (40 tons) of great strong wood/ha.

Neem wood is hard and moderately substantial and strict symbols in certain pieces of India. The wood seasons well, with the exception of end parting. Being sturdy and termite safe, neem wood is utilized in making wall posts, shafts for house development, furniture and so on there is developing business sector in some European nations for light-shaded neem wood for making family furniture. Post wood is particularly significant in agricultural nations; the tree's capacity to resprout subsequent to slicing and to regrow its overhang in the wake of pollarding makes it exceptionally fit to shaft creation (National Research Council, 1992). Neem develops quickly and is a decent wellspring of kindling and fills; the charcoal has high calorific worth.

External and Internal Use

In the history of mankind, medicinal plants have used for ages. In medicinal chemistry, such kinds of natural organic products. The external use of neem treats the following disorders.

- Psoriasis
- inflammatory condition
- infected wounds
- abscesses
- sinusitis
- lopecia
- gout

The internal use of neem fixes the disorders like:

- malaria
- filaria
- spleenomeglay
- respiratory disorders
- smallpox
- Vaginal Disorders
- AIDS

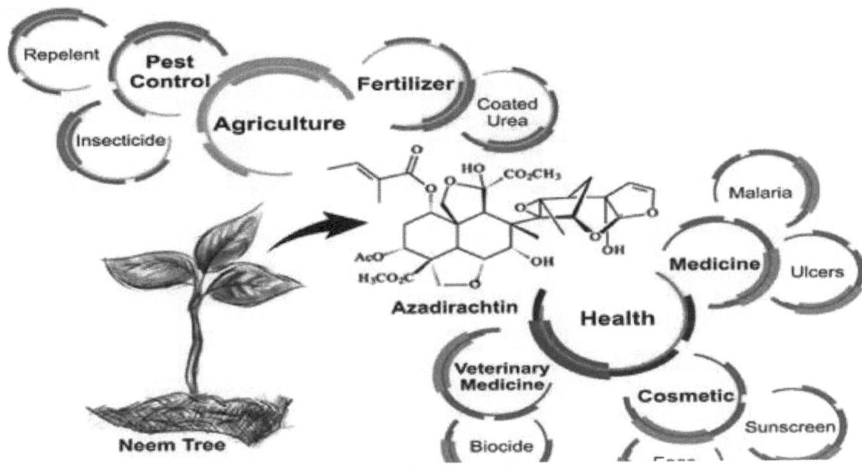

Figure 10: Applications or uses of Neem tree in different departments.

Dental

By the studies, it showed that neem helps in prohibiting the production of plaque and gingivitis. For the protection against cavity and periodontal disease, neem is advantageous. The neem mouthwash is an alternate to chlorhexidine gluconate treatments. This is considered much effective and economical. Researchers concluded that *A. indica* rinse is comparatively much effective than chlorhexidine to reduce the periodontal indices. The petroleum ether and chloroform extraction possess strong anti-microbial activities against the *S. mutans*. The S. mutans *Streptococcus salivarius* was highly resisted by chloroform extract.

Dandruff

Though the process for removing dandruff is still not clear, it is proved that it removes out dandruff. The neem oil comprises the characteristics like anti-fungal and anti-bacteria which clear away dandruff.

Therapeutical and Medicinal uses of Neem

Neem play a pivotal role in battling the disease particularly by activating the anti-oxidative enzyme.

a) Anti-oxidant activity:

In the era of diseases reacting oxygen comes to the top of the list. But neutralizing the free radical is considered as a crucial step for preventing any disease. Prior to attack in cells, free radicals are inactivated by anti-oxidants. Following is the role it is involved in.

- Activating the anti-oxidative enzyme that is involved in maintaining the damage which is involved in free radicals.

The antioxidant activity is found in many medicinal plant leaves, fruits, seeds, bark and roots. The extracts of neem and leaf were studied and it concludes that they have rich amount of antioxidant properties. Another experiment was made on Siamese neem tree parts and stem bark extracts to study the antioxidant. This concluded that their extraction has greater potential of antioxidant. An in vitro condition was applied to assess the antioxidant activity. Following is the result which clearly shows that chloroform crude extraction of neem has greater amount of anti-oxidant.

Chloroform > butanol > ethyl acetate extract > hexane extract > methanol extract.

The aqueous extract that was obtained from the Siamese neem tree leaves and stem ethanol extraction has greater amount of free radical which displays 50% scavenging activity at 30.6 mg/mL.

b) Anti- Cancerous activity

Cancer is regarded as the most dominant health issue in the world. The mutation caused in the body leads to cancer in the body. The module-based treatment has advantages as well as disadvantages. Many studies showed that plants and elements have inhibitory effects on malignant cells. This occurs through apoptosis and many other molecular ways. Neem comprises flavonoids that play a pivotal part in inhibiting the development of cancer. It was concluded from the epidemiological that intake of high flavonoids can lead you towards less challenge caused by cancer. Neem limonoids is present in neem oil which are involved in preventing the mutagenic effects of 7, 12-dimethylbenz (a) anthracene. *Azadirachta indica* plays a crucial part in treating cancer. The accurate and authentic mechanism of molecular is not acknowledged. But the experiments showed that neem is involved in modulating signalling pathways of different types of cells. The neem contains many active ingredients which activate the suppressor genes of the tumour. It also inactivates the many genes that are involved in the formation of cancer i.e. VEGF, NF. *Azadirachta indica* is considered an exceptional activator of suppressor genes of the tumour. It also inhibits the phosphoinositol PI3K/Akt. *Azadirachta indica* is involved in many pathways which prevent malignancies.

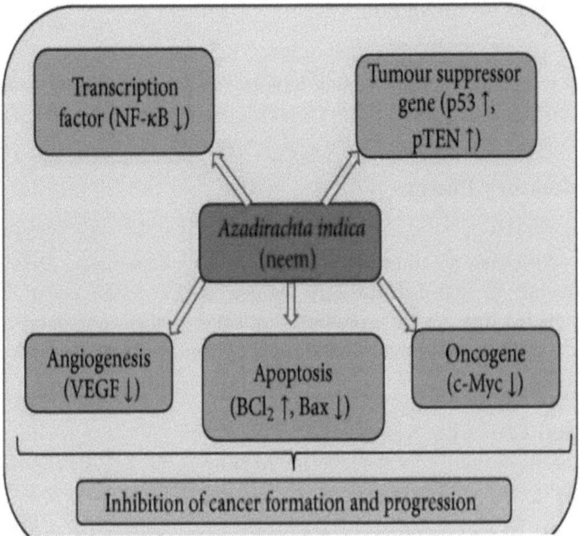

Figure 11 Anti- cancerous activity of Neem

c) Neem effect on Suppressor genes of tumor

The most important suppressing gene is p53 which restricts the production of abnormal cells. Researchers concluded that pro-apoptotic genes and some proteins like p53, Bcl-2-associated X protein (Bax) were regulated by the EFNL. This was involved in embellishing the expression like caspase-8 and caspase-3.

Nimbolide deregulates the proteins (mainly for surviving of cell) like:

- I-FLICE
- cIAP-1
- cIAP-2
- Bcl-2
- Bcl-xL
- Survivin

I. Neem effect on oncogene

An oncogene is responsible for developing tumors. An experiment was made to inspect the leaf extract c-Myc oncogene expression in 4T1 breast cancer BALB/c mice. The result concluded that the neem extract group suppresses the c-Myc oncogene expression.

II. Inflammatory Power

Plant and their compounds are utilized to perform as anti-inflammatory. 200mg dose of *A. indica* proved the presence of anti-inflammatory actions in the cotton pellet in rats. Many other researchers concluded that it is less productive than any other compound like dexamethasone. Moreover, the study concluded that nimbidin lowers and restricts the working of macrophages and neutrophils

Anti-Microbial Effect by Neem
Neem and its constituents are considered as the fighting agents against microbes, viruses and pathogens etc.

a) Antibacterial Activity

An evaluation was made to conclude that herbal supplements are best in resisting the microbial effects. It is clearly shown that they possess of large ratio of inhibition. The sodium hypochlorite is just 3% by ratio. Neem extraction and guava possess the compound which is used for the anti-bacterial properties. These are useful in controlling foodborne pathogens. It also concludes that extraction from seed and fruit comprises the anti-bacterial actions within a large number of concentrations.

b) Antiviral activity

It was concluded from the experiments that extraction from neem bark blocks the HSV-1. Moreover, the blocking actions were taken into notice when extraction was put to pre-incubation with the virus. Leaves extraction shows virus activities contrary to coxsackievirus virus B-4.

c) Antifungal Activity
To inspect the efficacy of neem leaf, an experiment was made which concluded that neem leaf extracts have the potential and ability to inhibit the growth of Aspergillus and Rhizopus. The extraction made of an alcoholic was more efficient for inhibiting the fungi species. While methanol and ethanol extract is involved in inhibiting the *Aspergillus flavus, Alternaria solani,* and *Cladosporium.*

d) Anti- Diabetic Actions:

An experiment was made to analyse the hypoglycemic activities of neem in rats (diabetic). The result showed that neem extraction which is 250mg/kg shows the level of glucose comparatively less than others. Leaves extraction of neem can naturally treat the disease like diabetes mellitus.

e) Anti-malarial activity

Plasmodium berghei was injected in albino mice. The result was concluded that extraction from leaf and neem reduce parasitemia by 51-80%. Further experiments resulted that Azadirachtin and limonoids which are present in the neem extraction are purely active on the vectors of malaria.

f) Healing of Wound:

Many plants and their parts are involved in healing the wound. *A. indica* extraction of leaves was used and it showed the healing actions in excision and incision wounds. The tensile strength was found higher than the control group. With the aid of inflammatory response and neovascularization, the neem could help out in healing the wounds.

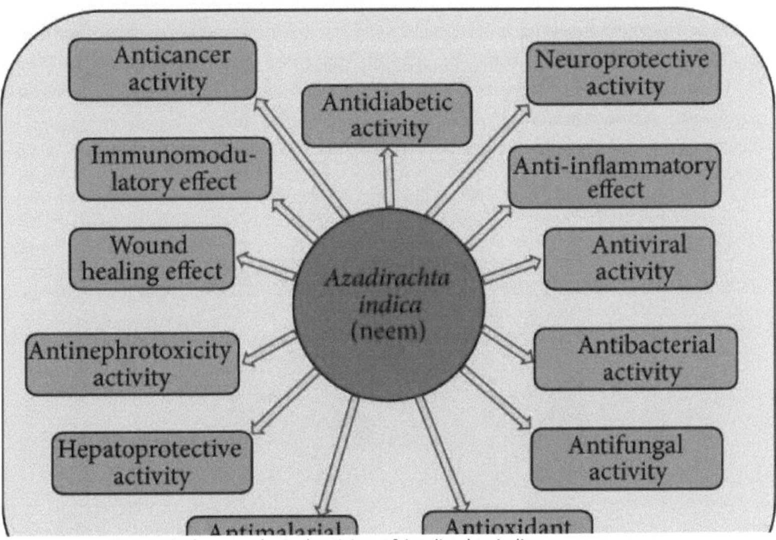

Figure 12 Pharmacological activites of Azadirachta indica

Side Effects of neem

Though the products are extracted from nature, they are still considered inefficient for humans. Its seed constitutes any type of fatty acid.

- The bark of the neem is considered safe for adults. The limit of the dose should not extend to 60 mg daily. This might can affect your kidneys as well as liver.
- You can use neem oil and cream onto the skin for 2 weeks which is probably safe.
- The extract of the neem gel is considered safe when used in the mouth for 6 weeks.

Precautions must be applied while using the neem orally. Many infants experienced the use of neem oil as poison and the doses were around 5–30mL. No clue of poison was found in the case of animals.

Threats to Neem

The neem tree arises from the dry forests of North East India. Its seed and leaves are used to extract an organic insecticide. Financial stability is the main factor in Northern Australia which has failed in planting the neem tree. An individual tree can produce up to 50000 seeds per year. Birds usually ingest the fruits and their seeds are spread by them to the other disease. Suckers are produced by them which blocks the vegetation. Neem is acknowledged as Class B and C weed in the northern territory region. This acknowledgement shows its legislations on buying, import and export of seeds.

Figure 13: Threats to Neem tree

Neem Tree Care

Neem Plant needs a lot of sunlight. They are not suitable to grow in excess water which makes them intolerant. Before planting the tree, the Soil must be dry. They can grow best at up to 120 degrees while in the pleasant weather they shed their leaves at below 5C. This tree doesn't grow in drought, extremely rainy areas and low temperatures. An appropriate amount of fertilizers, temperature and sunlight makes them elegant.

Figure 14: Neem Tree Care

Biotechnology Contribution

Each part of the plant especially leaves, bark and seeds are considered most important. Their roots can carry water and other nutrients from the deep ground. It could be the source of producing many fruitful products. The neem wood is used for manufacturing furniture and other domestic tools. Neem wood resembles teak wood in the terms of strength and resistance to fungicide and pesticide. This is enduring outdoors. There is a great gene diversity in the size of trees, fruit and morphology. There is reasonable uncertainty in the seeds, regardless of the habitat. Advancements in the production of azadirachtin can be carried out by the technique of clonal propagation. The conventional methods might seem achievable but are quite difficult for vegetative propagation of neem.

Thus, it is growing from seeds. With the plant cell and tissue culture technique, these limitations will behold in check and the manufacturing of clonal material will occur promptly. For the throughout the production of neem metabolites despite the season, an alternative was proposed.

This shows many favoured circumstances against the conventional methods.

- Multiplying rate occurs rapidly and produces a whole year. Thus, it produces thousands of plants in a year, just from a single piece of tissue.

- Multiplying rate decreases between a collection of trees and producing sufficient materials for the trials in the field.

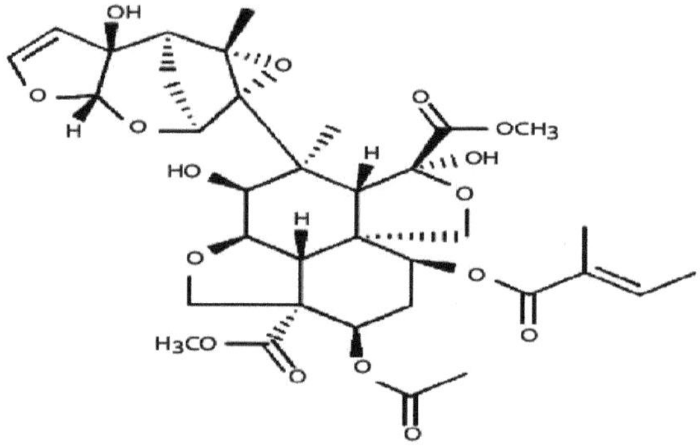

Figure 15 Chemical Structure of Azadirachtin

Thus, it is growing from seeds. With the plant cell and tissue culture technique, these limitations will be hold in check and the manufacturing of clonal material will occur promptly. For the throughout production of neem metabolites despite the season, an alternative was proposed.

Table 3: Regeneration in Tissue Cultures of Azadirachta indica

Sr.no	Modes of Propagation	Explant Used	Adult/Juvenile	Observations

1.	**Axillary Shoot Proliferation**	Nodal Segments	J	Axillary shoots → Plantlets
			A	Axillary shoots → Plantlets
		Apical & Axillary shoot bud	A	Axillary shoots → Plantlets
		Nodal segments from crown branches	A	Axillary shoots → did not survive
		Nodal segments from basal sprouts leaf	A	Axillary shoots → Plantlets
2.	**Triploid Production**	Immature endosperm	A	Callus→ Shoots→ Plantlets
3.	**Protoplast Culture**	Protoplast	A	Cell division and multiplication
4.	**Somatic Embryogenesis**	Roots Cotyledons	J J	Adv shoots→ Plantlets Calli→ Embryos→ Shoots Calli→ Embryos→ Plantlet

IN VITRO INSPECTION IN NEEM

✓ Axillary Shoot Proliferation

The formation of Shoot proliferation by the help of axillary buds is favorable where the desired products are mostly clonal plants. In this method the nodal segments are very useful. The reasonable access to clonal propagation is micro propagation through the induction of stems from already existing meristems. The apex shots cells are all diploid and under the specific culture circumstances show less genotypic changes. It shows the conservation of plant source properties.

Figure 16 Axillary Buds

This shows many favored circumstances against the conventional methods.

- Multiplying rate occurs rapidly and produce whole year. Thus, it produces thousands of plants in a year, just from a single piece of tissue.

- Multiplying rate decreases between collections of trees and producing sufficient materials for the trials in field.

✓ Adventitious Shoot Proliferation

This process is multistep and comprises continuous events, which is said as induction. The technique of micro propagation through the adventitious shoot happens directly and indirectly which happens by intervening in the phase of callus. The approach becomes less favourable for large scale clonal when indirect regeneration appears in somaclonal variation. Some hindrances were caused due to the representation of many genotypes and these were linked with the immature zygotic embryo. At this point, the prediction of seeds giving rise to an elite tree might be difficult. The technique of micro propagation by the de novo of shoot buds is significantly more favourable. Here the initial explant will not be considered as a limiting condition.

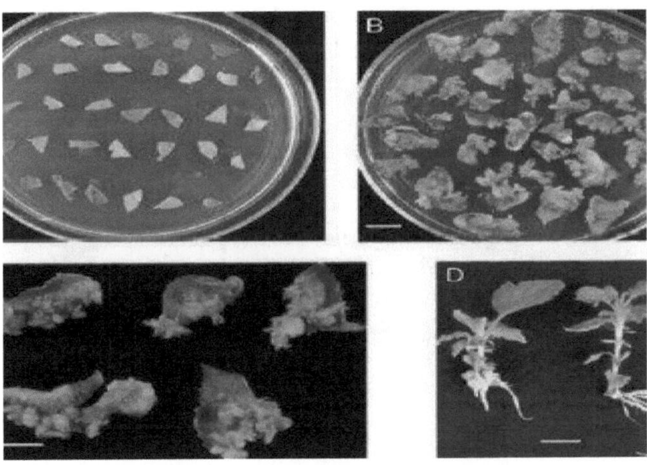

Figure 17 Shoot regeneration from cultured leaves

✓ Somatic Embryogenesis

In the tissue, the regeneration of plants may occur through organogenesis or somatic embryogenesis tissue culture. The latter shows many advantageous properties. The origin of Single-cell from embryos desire these processes for the efficient transformation of plants. Explant has various cells and tissues but just a few are capable of embryogenic induction. The tissue is specialized for embryogenic and gives a response much better to the induction treatment. Many changes are formed at the explants molecular and physiological stages. By the PGR treatment, the explant goes under reprogramming.

To obtain the response of an embryogenic, type of explant may be a serious factor. The best explant is considered as a zygotic embryo that is in an immature state. Its modern aspects showed favourable somatic embryogenesis. The hypocotyl area is considered as an alternative. To study the neem the considerable explant is immature zygotic embryos that resulted in higher responses of embryogenic in few days only. At the dicotyledons stage, the premature embryo showed its effectiveness in induced somatic embryos. The zygotic embryo which was in premature form was utilized by plants to get somatic embryogenesis like Castana Sativa Mill. The explants work efficiently metabolically and biochemically because of the availability of milky white ingredient found in cotyledons of neem. Neomorphs were declared different from the torpedo stage.

Neem is used all around the world and it is known for its numerous benefits. This plant is responsible for raising the economy of the country. Scientists have upgraded the value of neem by plant tissue culture and technique. The rising of somatic embryos arises the common stage for many other tissue culture techniques. It is considered very useful in research and industry.

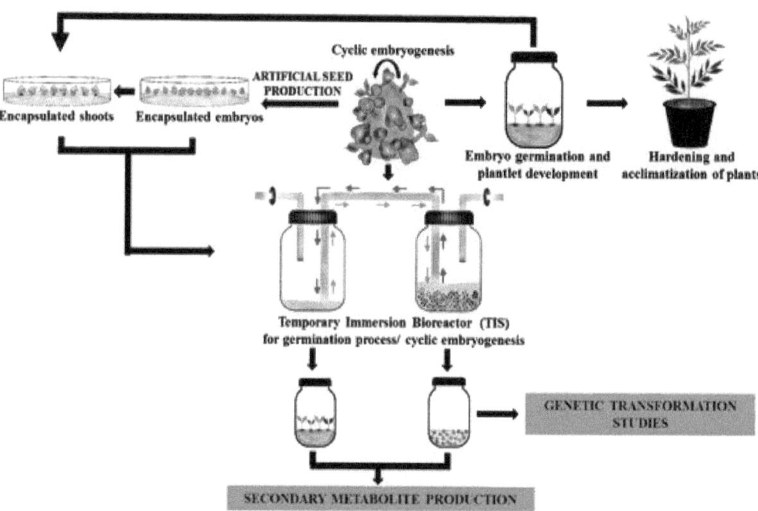

Figure 18 Somatic Embryogenesis in Neem

✓ **Triploid Production:**

Endosperm is considered a unique tissue. It is formed by double fertilization but it forms into the formless tissue. Lampe and Mills in 1933 tried to develop endosperm tissue. LaRue in 949 developed the first tissue culture of premature maize endosperm. Johri and Bhojwani declared about their development of the first totipotency of the endosperm cells. **Chaturvedi et al** reports the production of triploids from neem endosperm.

The property like a seed- sterile is found in triploid plants. But there are many cases where seed lessness from triploid favours the circumstances. Triploids are used commercially in the crops like apples, bananas and sugar beet. Azadirachtin is attained in kernels. Thus, it shows that triploids choose for tetratriterpenoid might not be acceptable. The production of triploids by joining induced superior tetraploids and diploids is a conventional method. Endosperm can regenerate the pant and this is requisite for the neem. By the technique of micropropagation, the specific triploids can be joined together.

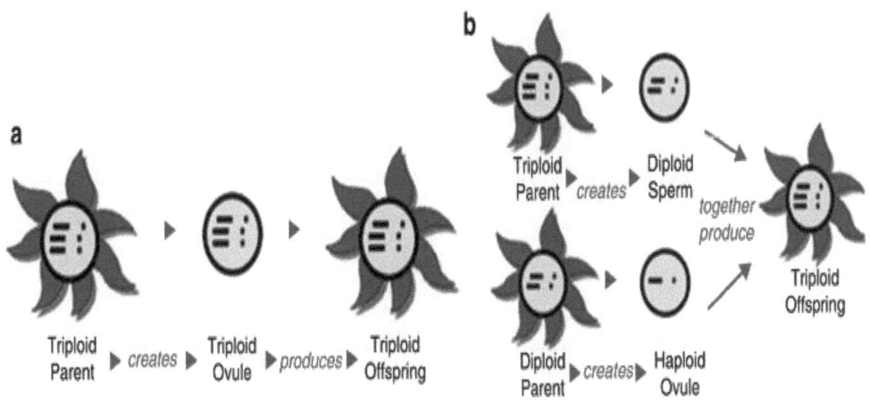

Figure 19 Applications of Triploid

Genetic Transformation

With the advanced technology, humans are now capable of changing the history of mankind. With the help of genetic engineering, we can study every gene in plant. This method is far better than the conventional ones because we can insert the desired gene into the plant. That is how we get the desired product from that plant. Many few changes have been made in the Neem plant till now. Tumors were formed that made distinctive shoots on basal medium. They were developed due to the seedlings which were infected with agrobacterium tumefactions. The shoots which were manufactured by the process formed octopine and they found resistant to kanamycin and formed hairy roots from stem and neem explants. This focused in inspecting the manufacturing of insect anti-feedant compounds particularly in neem.

Root cultures grow at extremely fast rate and resulted in 100 fold increase in 4 week, in the biomass. A distinctive result showed against the *Schistocerca gregaria*. In another observation, the seedlings were contaminated with Agrobacterium rhizogenes to attain the hair roots. Effect on various cultures were also studied. Different types of media were tested and Ohyama and Nitsch's basal medium developed large number of azadirachtin (0.017% DCW).

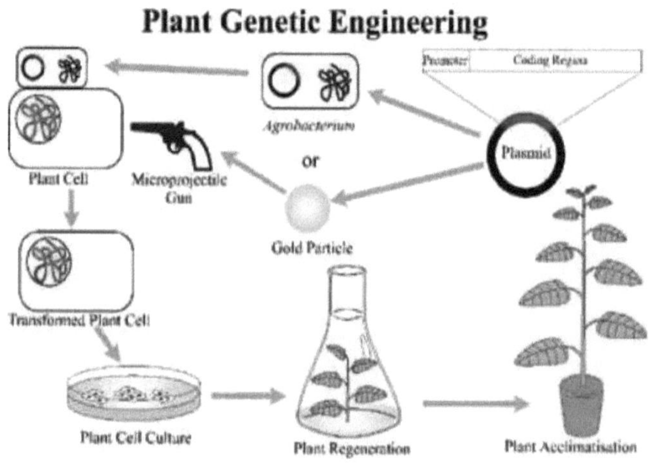

Figure 20 Schematic Diagram of Plant Genetic Transformation

Economic Strength by Neem

Due to its large properties, the neem tree is gaining value day by day. Its demand has increased in the areas like agriculture, cosmetics and many other industries. Neem oil is used for the products like facials creams and soaps etc. The oil, neem pesticides sell like hot cake in the agricultural sector due to their eco-friendly environment. By applying the urea with neem lessens the loss of nitrogen. A country like India reduces their subsidy to 6500Rs crore, just by producing 100% neem layered urea. African countries, Pakistan and India are enriched with this miracle. With the aid of proper knowledge and investment, they can raise their economy significantly.

Economic benefits of neem

The Neem tree is currently acquiring the significance because of its wide extent of commercialization in the space of agribusiness, veterinary, beauty care products, medication, toiletries and different ventures. Neem is presently turning into a mainstream in beauty care products and excellence help. A few organizations are presently utilizing Neem items (Neem oil and leaves) for creation of beauty care products like facial creams, nail shines, nail oils, shampoos, conditioners and so forth. The interest of Neem items are expanding step by step. Agribusiness area is currently turning into a significant buyer of the neem items viz. Neem oil, neem cake and neem based pesticides. Being eco-accommodating and regular wellspring of phyto-synthetic substances and supplements it is liked to apply Neem excrement and pesticide in horticulture particularly in natural cultivating everywhere. The covering of urea with Neem items has been offered need to limit the misfortunes of nitrogen. Legislature of India has as of late permitted compost firms to deliver 100% Neem Coated Urea – a move pointed toward assisting ranchers with boosting pay and lessening endowment bill by up to Rs. 6,500 crore. Legislature of India has discarded the cap on Neem – covered Urea and presently it very well may be delivered 100%. It is a success – win circumstance for every one of the major parts in the field-the business, ranchers and Neem seed assortment Agencies.

The nations like India and numerous African nations having enormous number of neem trees are the significant wellspring of neem foods grown from the ground items. In the event that business manor and agro-ranger service including neem is advocated, the potential goes up altogether, with positive and huge externalities for pesticides, manures, domesticated animals, dairying and other worth added items.

Neem tree can possibly help little and minor ranchers in provincial India, Africa and Latin America. Ranchers, who have restricted assets, can benefit from numerous points of view from neem. There are effectively exploitable, business and pay age open doors in the development of neem and preparing of neem items, some of which are conceivable in a decentralized way based on little ventures. The greater part of the non-industrial nations have tremendous regions under negligible grounds with low efficiency. As neem has various utilizations, its harvest on peripheral grounds can make a critical commitment to rustic economies.

Impetuses for neem seed assortment with regards to the current monetary real factors should be fortified alongside hierarchical enhancements for advertising of neem seeds. Authoritative, monetary sources of info and a strategy for coordinating neem in the system of agribusiness, provincial and little ventures strategies is required to understand this potential. Satisfactory stockpile of good quality neem seeds in an ideal design is basic for the business achievement of Neem. In India, offices as of now exist for extraction of oil from neem seeds. It is feasible to utilize the previous offices for getting severe concentrates. Nonetheless, not at all like on account of oil extraction, to get great quality Neem dynamic concentrates, the extraction methodology must be of elevated expectations. Appropriate consideration must be taken in taking care of seeds at different stages including acquisition, drying and capacity.

Future of Neem

Nowadays the biggest challenge with agriculture is the production of food for billions of people. Scientists are working to remove global hunger without damaging nature. To control pesticides, various types of chemicals are involved which increases environmental pollution. The utilization of neem oil has been discussed for the production of healthy agriculture and the environment.

The introduction of neem in the agricultural sector is not for the first time. In India, extraction of neem is used for pesticides. According to the researchers, it has been proved that neem is safe for the workers. it can also be used the whole year. Neem has been declared as the soil and crop improving agent. The organic and inorganic materials present in the compounds flourish the quality and quantity of crops. Urea is considered as the basic need of nitrogen fertilizer and it is used worldwide for crops. Neem refrains the nitrification which aids the bacterial activity to slow down. That's how urea is saved in soil. Such properties make the neem sustainable and the crops are free from agrochemicals. This technique helps in solving the production of soil, but still, there are some conditions through which farmers have to be passed.

Biological control can be stated as the activity of natural agents which protects the crops from harmful agents which lead towards the damage of crops. They have been recognizing when the Chinese utilized the ants to maintain the pests around citrus in the 3rd century. Neem oil has been widely used for pest control. These natural products are somehow considered safer for managing the pests but they must be evaluated in IPM. Experiment was done to evaluate the efficiency of neem products. Colonies of Beauveria bassiana and Lecanicillium lecanii were compatible with many other products.

With the advancements in technology, new emerging field nanotechnology acts as a piece of novel equipment in the sector of agriculture. The nanotechnology system can flourish the efficiency of active ingredients by encapsulating the agrochemicals. This decreases the toxicity in the environment. This also lessens the volatilization and photobleaching etc. The utilization of nanoparticles administers the improved ways of conserving the neem oil which results in continuity on the target of pests. Still, some challenges must be resolved for the efficient working of nanoparticles linked with the insecticides. The challenges require the following solutions.

- Monitoring of the utilization of nanoparticles in the agriculture sector
- Scalable Nano formulations
- A complete study of nano pesticides
- Assessment of toxicity against the non-target species

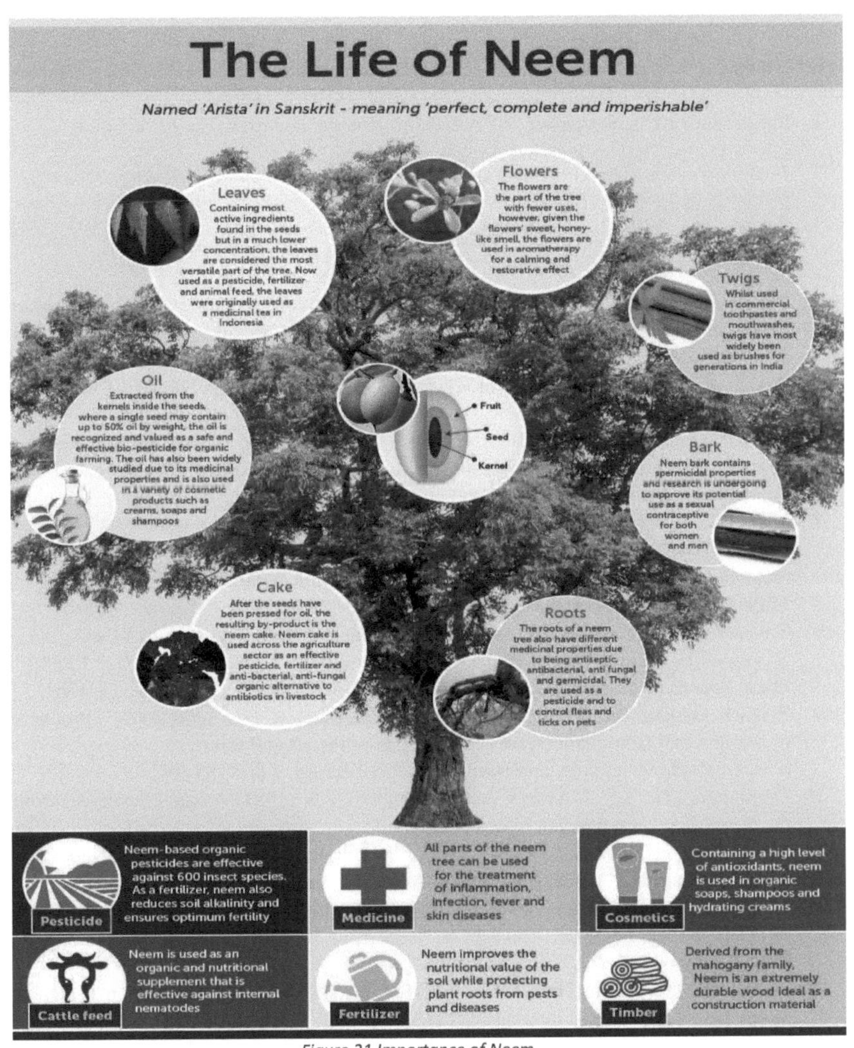

Figure 21 Importance of Neem

References

1. Sangeetha. (2020, November 26). *NEEM oil benefits and uses*. The Little Shine. https://thelittleshine.com/neem-oil-benefits-uses/.
2. *Patent on neem*. Neem Foundation. (2017, July 7). https://neemfoundation.org/about-neem/patent-on-neem/.
3. Sangeetha. (2020). *NEEM oil benefits and uses*. The Little Shine. https://thelittleshine.com/neem-oil-benefits-uses/.

4. "Azadirachta indica". Germplasm Resources Information Network (GRIN). Agricultural Research Service (ARS), United States Department of Agriculture (USDA). Retrieved 9 June 2017.
5. Bakshi, B.K. (1976). Forest Pathology: Principles and Practice in Forestry. Controller of Publications, Delhi.
6. Sidhu, O. P.; Kumar, Vishal; Behl, Hari M. (15 January 2003). "Variability in Neem (Azadirachta indica) with Respect to Azadirachtin Content". Journal of Agricultural and Food Chemistry. 51 (4): 910–915.
7. Narnoliya, L. K., Rajakani, R., Sangwan, N. S., Gupta, V., & Sangwan, R. S. (2014). Comparative transcripts profiling of fruit mesocarp and endocarp relevant to secondary metabolism by suppression subtractive hybridization in Azadirachta indica (neem). Molecular biology reports, 41(5), 3147–3162.
8. Srivastava, Smita; Srivastava, Ashok K. (17 August 2013). "Production of the Biopesticide Azadirachtin by Hairy Root Cultivation of Azadirachta indica in Liquid-Phase Bioreactors". Applied Biochemistry and Biotechnology. 171 (6): 1351–1361.
9. Prakash, Gunjan; Bhojwani, Sant S.; Srivastava, Ashok K. (1 August 2002). "Production of azadirachtin from plant tissue culture: State of the art and future prospects". Biotechnology and Bioprocess Engineering. 7 (4): 185–193.
10. Schmutterer, H., K.R.S. Ascher, and H. Rembold, eds. (1981). Natural Pesticides from the Neem Tree (Azadirachta indica A.Juss.).
11. Schmutterer, H. and K.R.S. Ascher, eds. (1984). Natural Pesticidesfrom the Neem Tree (Azadirachta indica A. Juss.) and Other Tropical Plants.
12. Ahmed, S. In press. Neem (Azadirachta indica) for pest control and rural development in Asia and the Pacific. Special session on neem from the 17th Pacific Science Congress, May 27 to June 2, 1991.
13. Benge, M.1986. Neem: The Cornucopia Tree. S&T/FENR Agro-Forestation Technical Series No. 5. Agency for International Development, Washington, D.C.
14. Jacobson, M., ed. 1988. 1988 Focus on Phytochemical Pesticides: Volume 1, the Neem Tree.CRC Press, Inc., and Boca Raton, Florida, USA.
15. Schmutterer, H.1990. Observations on pests of Azadirachta indica (neem tree) and of some Melia species. Journal of Applied Entomology 109:390-400.

16. Sanguanpong, U. and H. Schmutterer. In press. Laboratory trials on the effects of neem oil and neem-seed extracts against the two-spotted spider mite Tetranychus urticae Koch. Zeitschrift für Pflanzenkrankheiten und Pflanzenschutz.
17. Schmutterer, H.1984. Neem research in the Federal Republic of Germany since the first international neem conference.
18. Singh KK. (Ed.). Neem, a treatise. IK International Pvt. Ltd 2009,488.
19. Stoll, G.1986. Natural Crop Protection, Based on Local Resources in the Tropics.Josef Margraf, Publisher, Aichtal, Germany. 186 pp.
20. Saxena, R.C.1989. Insecticides from neem. Pages 110-135 in Arnason et al., 1988
21. Badam, L., R.P. Deolankar, M.M. Kulkarni, B.A. Nagsampgi, and U.V. Wagh. 1987. In vitro antimalarial activity of neem (Azadirachta indica A. Juss) leaf and seed extracts. Indian Journal of Malariology24:11 1-117.
22. Elvin-Lewis, M.1980. Plants used for teeth cleaning throughout the world. Journal of Preventive Dentistry6:61-70.
23. Nath, K., D.K. Agrawal, Q.Z. Hasan, S.J. Daniel, and V.R.B. Sastry. 1989. Water-washed neem (Azadirachta indica) seed kernel cake in the feeding of milch cows. Animal Production48:497-502.
24. Patel, R.P. and B.M. Trivedi. 1962. The in-vitro antibacterial activity of some medicinal oils. Indian Journal of Medical Research50:218-222.
25. Anderson, D.M.W., A. Hendrie, and A.C. Munro. 1972. The amino acid composition of some plant gums. Phytochemistry11:579-580.
26. Radwanski, S.A. and G.E. Wickens. 1981. Vegetative fallows and potential value of the neem tree (Azadirachta indica) in the tropics. Economic Botany35(4):398-414.
27. Sarkar, M.S. and P.C. Datta. 1986. Biosynthesis of beta sitosterol in-vitro culture of Azadirachta indica cotyledon tissues. Indian Drugs 23 (8).
28. Ahmed, S., S. Bamofleh, and M. Munshi. 1989. Cultivation of neem (Azadirachta indica, Meliaceae) in Saudi Arabia. Economic Botany 43:35-38.
29. Ahmed, S.1990. Symposium on Natural Resources for a Sustainable Agriculture, proceedings of a meeting, New Delhi, February 6-10, 1990, ed. R.P. Singh. Indian Society of Agronomy, New Delhi.
30. Allan EJ, Eeswara JP, Jarvis AP, Mordue (Luntz) AJ, Morgan ED, Stuchbury T. 2002. Induction of hairy root cultures of Azadirachta indica A. Juss. And their production of azadirachtin and other important insect bioactive metabolites. Plant Cell Rep. 21: 374–379
31. Satdive RK, Fulzele DP, Eapen S. 2007. Enhanced production of azadirachtin by hairy root cultures of Azadirachta indica A. Juss by elicitation and media optimization. J Biotechnol. 128 (2): 281-289
32. Jalaluddin M, Rajasekaran UB, Paul S, Dhanya RS, Sudeep CB, Adarsh VJ. Comparative evaluation of neem mouthwash on plaque and gingivitis: a double-blind crossover study. J Contemp Dent Pract. 2017; 18(7):567-571. doi:10.5005/jp-journals-10024-2085
33. Bansal V, Gupta M, Bhaduri T, Shaikh SA, Sayed FR, Bansal V, Agrawal A. Assessment of antimicrobial effectiveness of neem and clove extract against Streptococcus mutans

and Candida albicans: an in vitro study. Niger Med J. 2019; 60(6):285-289. doi:10.4103/nmj.NMJ_20_19

34. Thengane S, Joshi M, Mascarenhas AF. 1995. Somatic embryogenesis in neem (Azadirachta indica). In: Jain SM, Gupta P, Newton R, editors. Somatic Embryogenesis in Woody Plants. Vol. II. Important Selected Plants. Dordrecht: Kluwer Academic Publishers, pp. 357-37

35. Ermel K, Pahlich E, Schmutterer H. 1984. Comparison of the azadirachtin content of neem seeds from ecotypes of Asian and African origin. In: Proc. 2nd Int. Neem Conf., Rauischholzhausen, pp. 91-93.

36. Ermel K, Pahlich E, Schmutterer H. 1987 Azadirachtin content of neem kernels from different geographical locations, and its dependence on temperature, relative humidity and light. In: Proc. 3rd Int. Neem Conf., Nairobi, pp. 171-184.

37. Ermel K. 1995. Azadirachtin content of neem seed kernels from different regions of the world. In: Schmutterer H, editor. The Neem Tree: Source of Unique Natural Products for Integrated Pest Management, Medicine, Industry and Other Purposes. Weinheim: VCH, pp. 222-230

38. Dogra PD, Thapliyal RC. 1996. Gene resources and breeding potential, In: Randhawa NS, Parmar BS, editors. Neem. New Delhi: New Age International Pvt. Ltd., pp. 27-32

39. Mohan Ram HY, Nair MNB. 1996. Botany, In: Randhawa NS, Parmar BS, editors, Neem. New Delhi: New Age International Pvt. Ltd., pp. 6-26

40. Allan EJ, Eeswara JP, Jarvis AP, Mordue (Luntz) AJ, Morgan ED, Stuchbury T. 2002. Induction of hairy root cultures of Azadirachta indica A. Juss. and their production of azadirachtin and other important insect bioactive metabolites. Plant Cell Rep. 21: 374–379

41. Rao KS, Venkateswara R. 1985. Tissue culture of forest trees: Clonal Multiplication of Eucalyptus grandis L. Plant Sci. 40: 51-55

42. Marcotrigiano M, Jagannathan L. 1988. Paulownia tomentosa Steud. 'Somaclonal Snowstorm'. Hort Sci. 23: 226-227.

43. Chaturvedi R, Razdan MK, Bhojwani SS (2004) In vitro morphogenesis in zygotic embryo

44. Rout GR (2005) In vitro somatic embryogenesis in callus cultures of Azadirachta indica A. Juss— a multipurpose tree. J For Res 10:263–26

45. Sezgin M, Dumanoğlu H (2014) Somatic embryogenesis and plant regeneration from immature cotyledons of European chestnut (Castanea sativa Mill.). In Vitro Cell Dev Biol Plant 50:58–68

46. Gairi A, Rashid A (2004) TDZ-induced somatic embryogenesis in non-responsive caryopses of Rice using a short treatment with 2, 4-D. Plant Cell Tiss Organ Cult 76:29–33

47. Gairi A, Rashid A (2005) Direct differentiation of somatic embryos on cotyledons of Azadirachta indica. Biol Plantarum 49:169–173

48. Chaturvedi R, Razdan MK, Bhojwani SS (2004) In vitro morphogenesis in zygotic embryo cultures of neem (Azadirachta indica A. Juss.). Plant Cell Rep 22:801–809

49. Bhojwani SS, Razdan MK. 1996. Plant Tissue Culture: Theory and Practice. Amsterdam: Elsevier, pp. 502

50. Chaturvedi R. 2003. Isolation and culture of Neem (Azadirachta indica A. Juss.) callus protoplast. Phytomorphology 53: 57-61.

51. Elliott FC. 1958. Plant Breeding and Cytogenetic. McGraw-Hill, NewYork, USA.

52. Naina NS, Gupta PK, Mascarenhas AF. 1989. Genetic transformation and regeneration of transgenic neem (Azadirachta indica) plants using Agrobacterium tumefaciens. Curr Sci. 58: 184-187

53. Nunes P. X., Silva S. F., Guedes R. J., Almeida S. Phytochemicals as Nutraceuticals—Global Approaches to Their Role in Nutrition and Health. InTech; 2012. Biological oxidations and antioxidant activity of natural products

54. Rahmani A. H., Aly S. M. Nigella sativa and its active constituent's thymoquinone shows pivotal role in the diseases prevention and treatment. Asian Journal of Pharmaceutical and Clinical Research. 2015;8(1):48–53.

55. Ghimeray A. K., Jin C. W., Ghimire B. K., Cho D. H. Antioxidant activity and quantitative estimation of azadirachtin and nimbin in Azadirachta indica A. Juss grown in foothills of Nepal. African Journal of Biotechnology. 2009;8(13):3084–3091.

56. Sithisarn P., Supabphol R., Gritsanapan W. Antioxidant activity of Siamese neem tree (VP 1209) Journal of Ethnopharmacology. 2005;99(1):109–112. doi: 10.1016/j.jep.2005.02.008.

57. Priyadarsini R. V., Manikandan P., Kumar G. H., Nagini S. The neem limonoids azadirachtin and nimbolide inhibit hamster cheek pouch carcinogenesis by modulating xenobiotic-metabolizing enzymes, DNA damage, antioxidants, invasion and angiogenesis. Free Radical Research. 2009;43(5):492–504. doi: 10.1080/10715760902870637.

58. Nahak G., Sahu R. K. Evaluation of antioxidant activity of flower and seed oil of Azadirachta indica A. juss. Journal of Applied and Natural Science. 2011;3(1):78–81.

59. Kiranmai M., Kumar M., Ibrahim M. Free radical scavenging activity of neem tree (Azadirachta indica A. Juss Var., Meliaceae) root barks extract. Asian Journal of Pharmaceutical and Clinical Research. 2011; 4:134–136.

60. Sithisarn P., Supabphol R., Gritsanapan W. Antioxidant activity of Siamese neem tree (VP1209) Journal of Ethnopharmacology. 2005;99(1):109–112. doi: 10.1016/j.jep.2005.02.008.

61. Rahmani A. H., Alzohairy M. A., Khan M. A., Aly S. M. Therapeutic implications of black seed and its constituent thymoquinone in the prevention of cancer through inactivation and activation of molecular pathways. Evidence-Based Complementary and Alternative Medicine. 2014; 2014:13. doi: 10.1155/2014/724658.724658

62. Le Marchand L. Cancer preventive effects of flavonoids—a review. Biomedicine and Pharmacotherapy. 2002; 56(6):296–301. doi: 10.1016/s0753-3322(02)00186-5.

63. Kumar G. H., Vidya Priyadarsini R., Vinothini G., Vidjaya Letchoumy P., Nagini S. The neem limonoids azadirachtin and nimbolide inhibit cell proliferation and induce apoptosis in an animal model of oral oncogenesis. Investigational New Drugs. 2010; 28(4):392–401. doi: 10.1007/s10637-009-9263-3

64. Harish Kumar G., Chandra Mohan K. V. P., Jagannadha Rao A., Nagini S. Nimbolide a limonoid from Azadirachta indica inhibits proliferation and induces apoptosis of human choriocarcinoma (BeWo) cells. Investigational New Drugs. 2009; 27(3):246–252. doi: 10.1007/s10637-008-9170-z

65. Arumugam A., Agullo P., Boopalan T., et al. Neem leaf extract inhibits mammary carcinogenesis by altering cell proliferation, apoptosis, and angiogenesis. Cancer Biology and Therapy. 2014; 15(1):26–34. doi: 10.4161/cbt.26604

66. Subapriya R., Kumaraguruparan R., Nagini S. Expression of PCNA, cytokeratin, Bcl-2 and p53 during chemoprevention of hamster buccal pouch carcinogenesis by ethanolic neem (Azadirachta indica) leaf extract. Clinical Biochemistry. 2006; 39(11):1080–1087. doi: 10.1016/j.clinbiochem.2006.06.013

67. Subapriya R., Bhuvaneswari V., Nagini S. Ethanolic neem (Azadirachta indica) leaf extract induces apoptosis in the hamster buccal pouch carcinogenesis model by modulation of Bcl-2, Bim, caspase 8 and caspase 3. Asian Pacific Journal of Cancer Prevention. 2005; 6(4):515–520.

68. Cohen E., Quistad G. B., Casida J. E. Cytotoxicity of nimbolide, epoxyazadiradione and other limonoids from neem insecticide. Life Sciences. 1996; 58(13):1075–1081. doi: 10.1016/0024-3205(96)00061-6.

69. Gupta S. C., Reuter S., Phromnoi K., et al. Nimbolide sensitizes human colon cancer cells to TRAIL through reactive oxygen species- and ERK-dependent up-regulation of death receptors, p53, and Bax. The Journal of Biological Chemistry. 2011; 286(2):1134–1146.

70. Li J., Yen C., Liaw D., et al. PTEN, a putative protein tyrosine phosphatase gene mutated in human brain, breast, and prostate cancer. Science. 1997; 275(5308):1943–1947.

71. Khan S., Kumagai T., Vora J., et al. PTEN promoter is methylated in a proportion of invasive breast cancers. International Journal of Cancer. 2004; 112(3):407–410.

72. Othman F., Motalleb G., Lam Tsuey Peng S., Rahmat A., Basri R., Pei Pei C. Effect of neem leaf extract (Azadirachta indica) on c-Myc oncogene expression in 4T1 breast cancer cells of BALB/c mice. Cell Journal. 2012; 14(1):53–60.

73. Chattopadhyay R. R. Possible biochemical mode of anti-inflammatory action of Azadirachta indica A. Juss. In rats. Indian Journal of Experimental Biology. 1998; 36(4):418–420.

74. Mosaddek A. S. M., Rashid M. M. U. A comparative study of the anti-inflammatory effect of aqueous extract of neem leaf and dexamethasone. Bangladesh Journal of Pharmacology. 2008; 3(1):44–47. doi: 10.3329/bjp.v3i1.836

75. Kaur G., Sarwar Alam M., Athar M. Nimbidin suppresses functions of macrophages and neutrophils: relevance to its anti-inflammatory mechanisms. Phytotherapy Research. 2004; 18(5):419–424. doi: 10.1002/ptr.1474.

76. Arora N., Koul A., Bansal M. P. Chemopreventive activity of Azadirachta indica on two-stage skin carcinogenesis in murine model. Phytotherapy Research. 2011; 25(3):408–416. doi: 10.1002/ptr.3280.

77. Biswas K., Chattopadhyay I., Banerjee R. K., Bandyopadhyay U. Biological activities and medicinal properties of Neem (Azadirachta indica) Current Science. 2002;82(11):1336–1345

78. Kumar S., Agrawal D., Patnaik J., Patnaik S. Analgesic effect of neem (Azadirachta indica) seed oil on albino rats. International Journal of Pharma and Bio Sciences. 2012;3(2):P222–P225

79. Ghonmode W. N., Balsaraf O. D., Tambe V. H., Saujanya K. P., Patil A. K., Kakde D. D. Comparison of the antibacterial efficiency of neem leaf extracts, grape seed extracts and 3% sodium hypochlorite against E. feacalis—an in vitro study. Journal of International Oral Health. 2013; 5(6):61–66.

80. Mahfuzul Hoque M. D., Bari M. L., Inatsu Y., Juneja V. K., Kawamoto S. Antibacterial activity of guava (Psidium guajava L.) and neem (Azadirachta indica A. Juss.) extracts against foodborne pathogens and spoilage bacteria. Foodborne Pathogens and Disease. 2007; 4(4):481–488. doi: 10.1089/fpd.2007.0040.

81. Yerima M. B., Jodi S. M., Oyinbo K., Maishanu H. M., Farouq A. A., Junaidu A. U. Effect of neem extracts (Azadirachta indica) on bacteria isolated from adult mouth. Journal of Basic and Applied Sciences. 2012; 20:64–67.

82. Tiwari V., Darmani N. A., Yue B. Y. J. T., Shukla D. In vitro antiviral activity of neem (Azardirachta indica L.) bark extract against herpes simplex virus type-1 infection. Phytotherapy Research. 2010; 24(8):1132–1140. doi: 10.1002/ptr.3085.

83. Badam L., Joshi S. P., Bedekar S. S. 'In vitro' antiviral activity of neem (Azadirachta indica. A. Juss) leaf extract against group B coxsackieviruses. Journal of Communicable Diseases. 1999;31(2):79–90.

84. Mondali N. K., Mojumdar A., Chatterje S. K., Banerjee A., Datta J. K., Gupta S. Antifungal activities and chemical characterization of Neem leaf extracts on the growth of some selected fungal species in vitro culture medium. Journal of Applied Sciences and Environmental Management. 2009;13(1):49–53

85. Anjali K., Ritesh K., Sudarshan M., Jaipal S. C., Kumar S. Antifungal efficacy of aqueous extracts of neem cake, karanj cake and vermicompost against some phytopathogenic fungi. The Bioscan. 2013; 8:671–674.

86. Shrivastava D. K., Swarnkar K. Antifungal activity of leaf extract of neem (Azadirachta indica Linn) International Journal of Current Microbiology and Applied Sciences. 2014;3(5):305–308 Natarajan V., Venugopal P. V., Menon T. Effect of Azadirachta indica (Neem) on the growth pattern of dermatophytes. Indian Journal of Medical Microbiology. 2003; 21(2):98–101.

87. Lloyd C. A. C., Menon T., Umamaheshwari K. Anticandidal activity of Azadirachta indica. Indian Journal of Pharmacology. 2005; 37(6):386–389. doi: 10.4103/0253-7613.19076.

88. Amadioha A. C., Obi V. I. Fungitoxic activity of extracts from Azadirachta indica and Xylopia aethiopica on Colletotrichum lindemuthianum in cowpea. Journal of Herbs, Spices and Medicinal Plants. 1998; 6(2):33–40. doi: 10.1300/j044v06n02_04. Jabeen K., Hanif S., Naz S., Iqbal S. Antifungal activity of Azadirachta indica against Alternaria solani . Journal of Life Sciences and Technologies. 2013;1(1):89–93. doi: 10.12720/jolst.1.1.89-93.

89. Dholi S. K., Raparla R., Mankala S. K., Nagappan K. Invivo antidiabetic evaluation of Neem leaf extract in alloxan induced rats. Journal of Applied Pharmaceutical Science. 2011;1(4):100–105.

90. Joshi B. N., Bhat M., Kothiwale S. K., Tirmale A. R., Bhargava S. Y. Antidiabetic properties of Azardiracta indica and Bougainvillea spectabilis: In vivo studies in murine diabetes model. Evidence-Based Complementary and Alternative Medicine. 2011;2011:9. doi: 10.1093/ecam/nep033.561625

91. Akter R., Mahabub-Uz-Zaman M., Rahman M. S., et al. Comparative studies on antidiabetic effect with phytochemical screening of Azadirachta indicia and Andrographis paniculata . IOSR Journal of Pharmacy and Biological Sciences. 2013;5(2):122–128. doi: 10.9790/3008-052122128.

92. Chatterjee A., Saluja M., Singh N., Kandwal A. To evaluate the antigingivitis and antipalque effect of an Azadirachta indica (neem) mouthrinse on plaque induced gingivitis: a double-blind, randomized, controlled trial. Journal of Indian Society of Periodontology. 2011;15(4):398–401. doi: 10.4103/0972-124x.92578.

93. Lekshmi N. C. J. P., Sowmia N., Viveka S., Brindha Jr., Jeeva S. The inhibiting effect of Azadirachta indica against dental pathogens. Asian Journal of Plant Science and Research. 2012;2(1):6–10

94. Chava V. R., Manjunath S. M., Rajanikanth A. V., Sridevi N. The efficacy of neem extract on four microorganisms responsible for causing dental caries viz Streptococcus mutans, Streptococcus salivarius, Streptococcus mitis and Streptococcus sanguis: an in vitro study. Journal of Contemporary Dental Practice. 2012;13(6):769–772. doi: 10.5005/jp-journals-10024-122.

95. kin-Osanaiya B. C., Nok A. J., Ibrahim S., et al. Antimalarial effect of Neem leaf and Neem stem bark extracts on plasmodium berghei infected in the pathology and treatment of malaria. International Journal of Research in Biochemistry and Biophysics. 2013;3(1):7–14.

96. Mulla M. S., Su T. Activity and biological effects of neem products against arthropods of medical and veterinary importance. Journal of the American Mosquito Control Association. 1999;15(2):133–152

97. Dhar R., Dawar H., Garg S., Basir S. F., Talwar G. P. Effect of volatiles from neem and other natural products on gonotrophic cycle and oviposition of Anopheles stephensi and An. culicifacies (Diptera: Culicidae) Journal of Medical Entomology. 1996;33(2):195–201. doi: 10.1093/jmedent/33.2.195.

98. Nathan S. S., Kalaivani K., Murugan K. Effects of neem limonoids on the malaria vector Anopheles stephensi Liston (Diptera: Culicidae) Acta Tropica. 2005;96(1):47–55. doi: 10.1016/j.actatropica.2005.07.002

99. Ofusori D. A., Falana B. A., Ofusori A. E., Abayomi T. A., Ajayi S. A., Ojo G. B. Gastroprotective effect of aqueous extract of neem Azadirachta indica on induced gastric lesion in rats. International Journal of Biological and Medical Research. 2010;1:219–222

100. Barua C. C., Talukdar A., Barua A. G., Chakraborty A., Sarma R. K., Bora R. S. Evaluation of the wound healing activity of methanolic extract of Azadirachta

Indica (Neem) and Tinospora cordifolia (Guduchi) in rats. Pharmacologyonline. 2010; 1:70–77.

101.   Osunwoke Emeka A., Olotu Emamoke J., Allison Theodore A., Onyekwere Julius C. The wound healing effects of aqueous leave extracts of azadirachta indica on wistar rats. Journal of Natural Science and Research. 2013;3(6)

yes
I want morebooks!

Buy your books fast and straightforward online - at one of world's fastest growing online book stores! Environmentally sound due to Print-on-Demand technologies.

Buy your books online at
www.morebooks.shop

Kaufen Sie Ihre Bücher schnell und unkompliziert online – auf einer der am schnellsten wachsenden Buchhandelsplattformen weltweit! Dank Print-On-Demand umwelt- und ressourcenschonend produzi ert.

Bücher schneller online kaufen
www.morebooks.shop

KS OmniScriptum Publishing
Brivibas gatve 197
LV-1039 Riga, Latvia
Telefax: +371 686 204 55

info@omniscriptum.com
www.omniscriptum.com

Printed by Books on Demand GmbH, Norderstedt / Germany